Clice de Araújo Mendonça
Carlos Alessandro Chioderoli
Mª Albertina Monteiro dos Reis

Mechanisation and soil physics in maize and Brachiaria intercropping

Clice de Araújo Mendonça
Carlos Alessandro Chioderoli
Mª Albertina Monteiro dos Reis

Mechanisation and soil physics in maize and Brachiaria intercropping

Consortium of maize with Brachiaria brizantha: analysis of soil physics and performance of the mechanised unit

ScienciaScripts

Imprint
Any brand names and product names mentioned in this book are subject to trademark, brand or patent protection and are trademarks or registered trademarks of their respective holders. The use of brand names, product names, common names, trade names, product descriptions etc. even without a particular marking in this work is in no way to be construed to mean that such names may be regarded as unrestricted in respect of trademark and brand protection legislation and could thus be used by anyone.

Cover image: www.ingimage.com

This book is a translation from the original published under ISBN 978-3-330-99627-4.

Publisher:
Sciencia Scripts
is a trademark of
Dodo Books Indian Ocean Ltd. and OmniScriptum S.R.L publishing group

120 High Road, East Finchley, London, N2 9ED, United Kingdom
Str. Armeneasca 28/1, office 1, Chisinau MD-2012, Republic of Moldova, Europe
Managing Directors: Ieva Konstantinova, Victoria Ursu
info@omniscriptum.com

Printed at: see last page
ISBN: 978-620-8-59057-4

To our parents, who are primarily responsible for our education, our civic, ethical and moral training.

WE DEDICATE

SUMMARY

The intercropping of annual crops with forage is a technology that allows production to be verticalised and can promote improvements in soil physics, associated with the greater volume of straw on the surface, as well as providing greater productivity, consequently greater profit and greater sustainability of the production system. The aim of this study was to evaluate the intercropping of Brachiaria brizantha with maize for silage production, using three forage sowing methods and two maize row spacings, as well as to assess the energy and operational performance of the tractor-seeder combination and identify possible changes in the soil's physical attributes; the work was carried out in the experimental area of the Department of Agricultural Engineering at the Federal University of Ceará. A randomised block design was used, with a factorial scheme (3x2) + 2 with four replications, totalling 32 experimental units. The treatments consisted of two spacings between rows of maize, 0.45 and 0.90 m, and three ways of sowing the forage plant: pre-sowing sowing of the maize, in the sowing row together with the base fertiliser and post-sowing sowing of the maize, as well as single maize at both spacings as a control. The energy and operational performance of the tractor-seeder combination was assessed, as well as the physical characteristics of the soil. The results led to the conclusion that the 0.45 m spacing between rows provides better soil structuring with an increase in soil porosity, and better soil physical characteristics.
operational performance of the tractor-seeder combination, with less slippage of the seeder wheels when compared to the 0.90 m spacing.

Key words: Direct sowing. Agricultural mechanisation. Intercropping.

Summary

CHAPTER 1

INTRODUCTION

Due to population growth, we are currently living in a scenario where there is a growing demand for food production, both in terms of quantity and quality. In the agricultural sector, this scenario requires the use of technologies that make it possible to verticalise productivity while maintaining soil, water and air quality in a context of agronomic, economic and socio-environmental sustainability.

Among the technologies capable of increasing productivity is the Crop-Livestock Integration System (CLIS) associated with no-till farming, the main advantages of which are that the soil is not disturbed, it is covered all year round, thus reducing the risk of erosion, the physical and biological conditions of the soil are improved and there is the possibility of greater productivity without the need to open up new areas for agricultural purposes.

Crop-livestock integration allows the area to be farmed all year round, generally using intercropping between annual crops and tropical forage plants, such as brachiaria, for example. The forage crop can be sown in various ways and at different times, it will establish itself and, after the annual crop has been harvested, it will develop and can be used to feed the animals and/or produce straw for direct sowing of the next crop. The aim of this procedure is to favour an increase in the supply of agricultural products at a lower cost, due to the synergism between farming and grazing, as well as diversifying producers' incomes.

The intercropping of annual crops and tropical forage has emerged as an alternative production system for the recovery of areas degraded by intensive agricultural activity, as well as for the formation of mulch for the no-till system. However, knowledge of the behaviour of the species in competition for production factors is of great importance for the successful formation of the pasture and the satisfactory productivity of the grain crop.

Maize is the annual crop most commonly used in intercropping with forage crops, due to the large amount of straw it produces, as well as being one of the most important cereals produced and consumed in the world, both for human consumption, biomass energy generation and animal feed. In silage production, the crop stands out for its yield of green mass per unit area and its nutritional qualities.

However, in order to be successful with SILP, the producer needs to make correct use of the mechanised equipment required to install and maintain the system, making performance evaluation of these equipment, especially seed drills and seed drills, an indispensable tool for producers looking to compete in the market, as productivity depends directly on the quality of the sowing process.

Various factors can interfere with the sowing operation, which can be related to the seed, soil, machine, climate and operator. When it comes to intercropping, the variation in the soil's physical characteristics associated with the greater volume of straw can also affect the performance of the tractor-seeder combination.

There are many studies that show the advantages of intercropping, but there is a lack of information on the behaviour of this agricultural production system in the semi-arid region of Brazil, which justifies studies to establish the most suitable form of intercropping for the region's soil and climate conditions, which will provide greater yields of green and/or dry matter for animal feed, straw for direct sowing, without compromising grain yields and which is economically viable for the producer.

In view of the above, the aim of this study was to evaluate the intercropping of Brachiaria brizantha with maize for silage production, using three forage sowing methods and two maize row spacings, as well as assessing the energy and operational performance of the tractor-seeder combination and identifying possible changes in the soil's physical attributes

CHAPTER 2

LITERATURE REVIEW

2.1 Physical attributes of the soil

The intensive use of areas for agricultural production, combined with improper soil management techniques, has caused degradation of the soil structure, negatively influencing plant development and predisposing the soil to degradation (STONE & GUIMARÃES, 2005).

This occupation and use of agricultural areas requires the adoption of conservationist soil management and use measures aimed at rationalising the use of machinery, providing greater soil conservation and promoting the economic and environmental sustainability of the agro-ecosystem. According to Mendonça et al. (2013), the main conservation-based technologies are the no-till farming system, crop rotation and crop-livestock integration.

According to Chioderoli et al. (2012), the crop-livestock integration system can improve the physical conditions of the soil due to the greater production of straw provided by the consortium, improving soil cover, favouring water infiltration, allowing greater exploration of the soil profile by the roots, reducing the erosion process and consequently maintaining the stability of the system.

The inclusion of forage crops in grain and/or forestry systems alters the physical and chemical properties of the soil, promoting changes in its quality; mainly by increasing aggregate stability, which results in an increase in soil macroporosity and infiltration capacity. These improvements occur due to the presence of straw and pasture roots, which contribute to an increase in the soil's carbon content (LOSS et al., 2011).

According to Stone et al. (2003), the root system of forage plants favours soil decompression and improves soil structure, allowing for greater infiltration of water and air, as well as improving the soil's physical, chemical and biological properties. Seidel et al. (2014) reported that forage plants increase aggregate stability, resulting in an increase in soil macroporosity.

The inclusion of pastures in the rotation of annual crops improves the physical quality of the soil due to the combination of three main effects: the absence of tillage during the pasture cycle, the presence of a dense root system acting as an aggregating agent and an increase in the activity of soil macrofauna (MARCHÃO, 2007).

Therefore, the intercropping of brachiaria with maize can be an important vegetative practice, especially for altering the structure of inadequately managed soils; this is the case of some areas cultivated in the no-till system, where crop rotation is not carried out and cultivation is done under sparse straw. Poorly managed no-till areas can be noticeable for their increased soil density, decreased macroporosity, restricted root development and reduced productivity (KLUTHCOUSKI; YOKOYAMA, 2003; DENARDIN et al., 2008).

Different soil physical attributes have been used to characterise the physical changes resulting from soil compaction. According to (Marchão et al., 2007), the most commonly used are: soil mechanical resistance to penetration, density, porosity and hydraulic conductivity due to their relative ease of determination and the low cost of obtaining measurements.

Compaction is defined as a process by which the rearrangement of aggregates occurs due to a reduction in the spaces between the aggregates, causing the particles to be closer together, increasing the density of the soil (Soil Science Society of America, 2010). Soil compaction is directly related to soil aggregation, since its spatial arrangement is altered both between and within aggregates (DEFOSSEZ & RICHARD, 2002).

According to Santos et al. (2009), the combination of perennial pastures followed by annual crops for grain production is more efficient in maintaining the soil's physical and chemical structure, favouring plant development, as species with different root systems and morphologies promote greater nutrient cycling and favour disease suppression.

Denardin et al. (2008) found that when maize was intercropped with Brachiaria brizantha, the density of an Eutrophic Red Latosol with a very clayey texture decreased in the 0-10 cm layer as a result of higher root mass production than in the lower 10-20 cm layer.

Rosa et al. (2012) assessed the impact of the time taken to adopt the no-till system on soil structure conditions in a Red Dystrophic Latosol and found that soil density and resistance to penetration increased as the time taken to adopt the no-till system increased.

Stone et al. (2002) report that in a no-till system, changes in soil structure are noticeable, such as an increase in density, a decrease in macroporosity, restrictions on root growth and the vertical movement of water as a result of the absence of tillage.

However, Conte et al. (2011) and Moreira et al. (2012) observed that there were no significant changes in soil density and porosity after seven and eight years, respectively, in an ILP-PD system. Other studies show that possible changes in the soil's physical structure do not alter the productivity of summer crops such as soya (FLORES et al., 2007) or the ideal conditions for plant production (SPERA et al., 2010).

Changes in soil structure resulting from compaction alter various characteristics of the porous system, affecting the size, distribution, continuity, volume and geometry of pores (LIMA et al., 2005).

Macropores are responsible for conducting O_2 and water through the profile, and the loss of these pores leads to a reduction in gas exchange (Camara; Klein, 2005; Genro Junior et al., 2009). Micropores, on the other hand, are responsible for retaining water in the soil, and with a reduction in pore size, water will be more strongly retained, making it more difficult for plants to absorb it (IMHOFF et al., 2001; BATEY, 2009).

Spera et al. (2009) evaluated the effect of grain production systems, annual winter pastures and perennial pastures under a no-till system, after ten years of cultivation, on the soil's physical characteristics. They found that compaction resulting from animal trampling interfered with the soil's physical attributes by increasing soil microporosity and decreasing macroporosity in the pasture phase, but without reaching levels capable of causing soil degradation.

The reduction in macroporosity caused by compaction impairs water infiltration, negatively affecting the soil's hydraulic conductivity. Blanco-Canqui et al. (2010) observed a reduction in saturated hydraulic conductivity, suggesting that compaction can reduce the infiltration of rainwater or the water applied during irrigation.

One of the main effects of integrating agricultural activities with livestock farming, according to Silva et al. (2011), is that changes occur in the soil's physical properties. Marchão et al. (2007), working on red latosol, concluded that after four years of grazing, compaction resulting from animal trampling in crop-livestock integration systems did not reach critical values that could limit subsequent annual crops.

These effects can be confirmed in the experiment conducted by Sales et al. (2010) on a Quatzarenic Neosol. The authors state that agricultural use of the soil causes changes in physical attributes, but these changes are not considered critical for plant root development.

According to Moreira et al. (2012), SILPs present complex interactions between soils, plants, machinery and animals, which can be translated into changes in soil structure, which makes it extremely important to quantify measurements that describe physical processes in soils, as the results contribute to understanding the impacts of management systems on the physical quality of soils.

2.2 Operational performance of the tractor-seeder combination

Operational analysis seeks to develop techniques for forecasting, planning, controlling and coordinating activities, with the aim of obtaining the maximum useful yield from all available resources, with the minimum expenditure (VALE, 2011).

The advantage of knowing the operational performance of an agricultural machine, or of mechanised combines, is to be able to rationalise mechanised agricultural operations. Rational mechanisation, according to Mialhe (1974), consists of using a system or set of machines in a technically and economically organised way to carry out the operations required for agricultural production.

Direct sowing techniques are a complement to current soil conservation techniques, which involve less tillage and removal of the soil and a greater amount of plant remains on the surface, with the advantage of reducing mechanisation operating costs, as well as the conservation aspects of reducing the soil's physical, chemical and biological characteristics (NAGAOKA & NOMURA, 2003).

In order to achieve success in a direct seeding system, the selection of suitable components for the seed drills is undoubtedly a necessary action to maintain the system's effectiveness. Among the items to be checked, their straw cutting mechanisms have a major influence on the operation (SILVA, 2000).

Therefore, in the direct sowing system, in which conventional soil preparation does not take place, the importance of a seeder that meets the basic precepts of cutting the plant mass and correctly depositing the seed and fertiliser (Chioderoli et al., 2010), covering the seeds with soil and straw and also compacting the soil laterally to the seed (Siqueira & Casão Júnior, 2004), is stressed in order to obtain better yields.

In conservation systems, such as direct seeding, the soil and cover conditions are generally less favourable to the deposition of seeds and fertilisers than those found in preparations with intense tillage, requiring greater care in this operation (Cortez, 2007). Therefore, the seed drills used in the direct seeding system must be robust and resistant, have efficient operating capacity and require the least amount of energy (LEVIEN et al., 2001).

According to Oliveira et al. (2000), in direct sowing, the performance of the seeder-soiler, in terms of cutting crop residues, opening the furrow and depositing seeds and fertiliser at suitable depths, deserves essential care to avoid problems in the initial phase of the crop.

The operational and energy performance of seeder-plough machines is usually assessed by working speed, fuel consumption, effective field capacity and wheel slip.

2.3 Travel speed and fuel consumption

Any operation with agricultural machinery must take into account the travelling speed of the agricultural implement, since this variable influences fuel consumption, traction demand and the quality of the service (SILVEIRA, 2008).

Travel speed alters tractor performance as it directly affects fuel consumption.

Filho et al. (2010) evaluated the fuel consumption of a tractor in three soil surface conditions as a function of working speed and observed that an increase in speed leads to higher fuel consumption regardless of the soil surface condition.

Silveira et al. (2013), when assessing the energy demand of a tractor- seed drill-fertiliser combination in a no-till system, as a function of travel speeds and engine shaft speeds, when sowing maize, found increases in hourly fuel consumption when varying travel speed and engine shaft speed. The authors also observed that each unit increase in travel speed and engine shaft speed resulted in an increase of 1.32 L h^{-1} and 0.0053 L h^{-1}, respectively, in fuel consumption.

With regard to sowing speed, Chioderoli et al. (2010) observed that there is variation in relation to soil preparation, with a higher value in no-till soil compared to scarified soil, associating the result with the greater amount of straw in the no-till areas, which contributes to efficient tyre contact with the soil, since much of the straw was made up of anchored fragments and therefore promoted better traction force, reflecting, directly, in the increase in speed.

Soil surface conditions also affect tractor fuel consumption. When assessing fuel consumption and operational field capacity when sowing oats in three soil managements, Mazetto et al. (2004) concluded that hourly fuel consumption and tractor slip were higher when sowing was carried out in a conventional system and with scarification, presenting greater tractor demands, since in these managements there was greater soil mobilisation when compared to the no-till system.

Mahl (2002) found an average 12% increase in hourly fuel consumption for each km h^{-1} increase in travelling speed during sowing operations. They also found an effect of travelling speed on the increase in hourly fuel consumption (OLIVEIRA, 1997), MAHL (2002) and MAHL et al. (2005),

For Borsatto (2005), the speed was higher (6.3 km h^{-1}) when the seeder was equipped with 4 rows spaced 0.90 m apart, compared to 7 rows spaced 0.45 m apart at a speed of 5.8 km h^{-1}. The 0.90 m spacing provided the lowest hourly consumption (12.9 L h^{-1}) and the lowest area consumption (5.66 L ha^{-1}) compared to the 0.45 m spacing, which had hourly consumption and area consumption of 13.35 L h^{-1} and 6.62

L ha$^{(-1)}$, respectively.

Jasper et al. (2015), working with two fertiliser furrowing mechanisms (Rod and Double Disc) and two row spacings in the maize crop (0.45 and 0.90 m), observed lower travel speed at the 0.45 m row spacing, a fact explained by Cepik (2010), who reports an increase in the traction force of the motor-mechanised unit due to the reduction in row spacing. At the reduced spacing, they also observed higher hourly fuel consumption.

According to Montanha et al. (2011), tractor fuel consumption is one of the highest costs in agricultural operations, and the total consumed is directly linked to factors such as the suitability and condition of the tractor-equipment combination, the depth of the operation, the type and condition of the soil, the total number of operations used in the soil preparation process, among others.

2.4 Operational Efficiency and Field Capacity

Field efficiency is an important criterion for verifying field capacity and for making important decisions about machine management (GRISSO et al., 1992).
According to ASAE standard D497.6 (2009), field efficiency is the ratio between the effective field capacity of a machine and its theoretical field capacity.

According to Silveira et al. (2006) the working or field capacity of agricultural machinery is a function of the following factors: the machine's working width, travel speed, the percentage of time it is stationary or not operated due to the time spent travelling to the area to be worked and manoeuvring, among other activities that may arise during the process.

Unnecessary travelling should be avoided when using agricultural equipment, because according to Simões et al. (2011), operational interruptions and corrective maintenance during operation can reduce the effective field yield of agricultural operations by approximately 16%.

Field capacity can be theoretical, effective or operational. Theoretical field capacity is obtained if the mechanised unit works 100% of the time at nominal speed, using 100% of its nominal width. According to Balastreire (1987), the effective field capacity (CcE) is obtained from the ratio between the actual performance of the machine (area worked) and the total field time, usually expressed in hectares per hour.

According to the American Society of Agricultural Engineers (1996) and Ribeiro (1997), the operational field capacity of a fertiliser seeder depends on the number of rows and the spacing between adjacent rows, the travel speed and the operational field efficiency.

According to Balastreire (1990), operational efficiency in the field for tractor seeders can vary from 50% to 85%, depending on the machine, the type and humidity of the soil, the amount and form of residue distribution, the shape of the crop, clutching, manoeuvring, the tractor driver's skill and autonomy. According to Silveira (2006), efficiency ranges from 60 to 80 per cent.

Rodrigues & Gamero (2006), comparing soil management systems and vegetation cover through effective field capacity, hourly and operational fuel consumption, found that the direct sowing system was the one with the best indices, regardless of the vegetation cover studied.

Rodrigues et al. (2011), studying the operational capacity of a mechanised combine as a function of different types of soil management and working speeds when sowing forage sorghum, identified better operational performance when working on soil under SPD and at a speed of 5 km h^{-1}. When evaluating a tractor-seeder combination when sowing maize, Mahl et al. (2004) found that increasing the speed at which the combination travelled from 4.4 km h^{-1} to 8.1 km h^{-1} resulted in an 86 % increase in effective field capacity.

Silveira et al. (2006) observed an average efficiency of 49.2% for the tractor-seeder-accumulator combination when sowing. The average operating speed was 3.4 km h^{-1} with a CV of 20.4%. The authors found that the operating speed varied greatly and the differences were due to the different soil conditions during sowing.

2.5 - Skating

Slippage is a relative movement in the direction of motion between the element providing traction and the surface when travelling, ASAE (1982). According to Correia et al. (1995), slip expressed as a percentage represents the loss of engine power through the wheelset and this phenomenon must be studied, quantified and reduced, but not completely eliminated.

In order for traction to occur, there must be skidding; however, if this exceeds certain limits, there can be a loss of grip and a reduction in wheel traction (Lanças & Upadhyaia, 1997). ASAE (1989) recommends that, in order to operate at maximum traction efficiency, the percentage of wheel slip should be: 8% to 10% on untilled soils, 11% to 13% on tilled soils and 14% to 16% on loose or sandy soils.

Wheel slip is one of the problems that can affect tractor performance, contributing to a reduction in tractive force and an increase in fuel consumption. In agricultural tractors, wheel slip occurs due to various factors, including tyre type, inflation pressure, wheel load, tread type and soil conditions, slope and soil water content (LANÇAS & UPADHYAIA, 1997).

Levien et al. (1999) found greater tractor wheel slip when sowing on ploughed and scarified soil compared to the no-till system. Chioderoli et al. (2010) also found that when sowing on scarified soil, the slip values were higher than those obtained for direct sowing. They associated this result with the fact that in no-till sowing, the roughness of the soil is maintained, resulting in less wheel spin, as the un-turned surface presents fewer obstacles to the wheels rolling.

Excessive wheel spin can lead to a loss of traction, heavy wear on the tyres with a reduction in their useful life, high fuel consumption, slower travel speed for the mechanised unit and, consequently, lower effective field capacity. It can also affect the quality of seed distribution.

The agricultural seeding machines available on the market have wheel-driven seed and fertiliser metering mechanisms whose efficiency is directly related to soil conditions, vertical load, tyre characteristics and the soil's physical properties. These factors determine the wheel slip rate, which is lower in direct seeding areas due to the lower resistance of the soil to rolling, allowing the seeding operation to be carried out with a higher water content in the soil compared to conventional tillage (VIEIRA & REIS, 2001).

Vale et al. (2008), when assessing the performance of a seeder in no-till and conventional tillage on Cambissolo soil, found 10.74 % and 7.18% slip for sowing in conventionally prepared soil and no-till, respectively. They justified this by saying that in conventional planting, because the soil is looser, it reduces the tyre's ability to adhere to the soil, resulting in greater slippage. In no-till, on the other hand, as the soil is not tilled, it can be considered firm and this helps to reduce skidding as it favours the tyre's adherence to the soil.

The opposite results were found by Mahl (2006), who showed that the straw on the soil surface makes it difficult for the tractor wheels to adhere to the ground, which can increase skidding and, consequently, cause a significant increase in the demand for traction force and power at the bar.

Furlani et al. (2004), when evaluating the operational performance of a precision fertiliser seeder in three soil preparation systems, observed that the slip of the drive wheels was lower when the tractor worked in unprepared soil, increasing significantly in conventional soil preparation and even more so in scarified soil, and observed the same behaviour for effective field capacity, with higher results in direct sowing.

CHAPTER 3

MATERIAL AND METHODS

3.1 - Characterisation of the study area

The experiment was conducted from December 2015 to March 2016 in an experimental mechanisation area belonging to the Department of Agricultural Engineering at the Federal University of Ceará (UFC Campus do Pici), located near the geographical coordinates: latitude 3°44'S, longitude 38°34'W Greenwich and altitude 19.5 m (Figure 1).

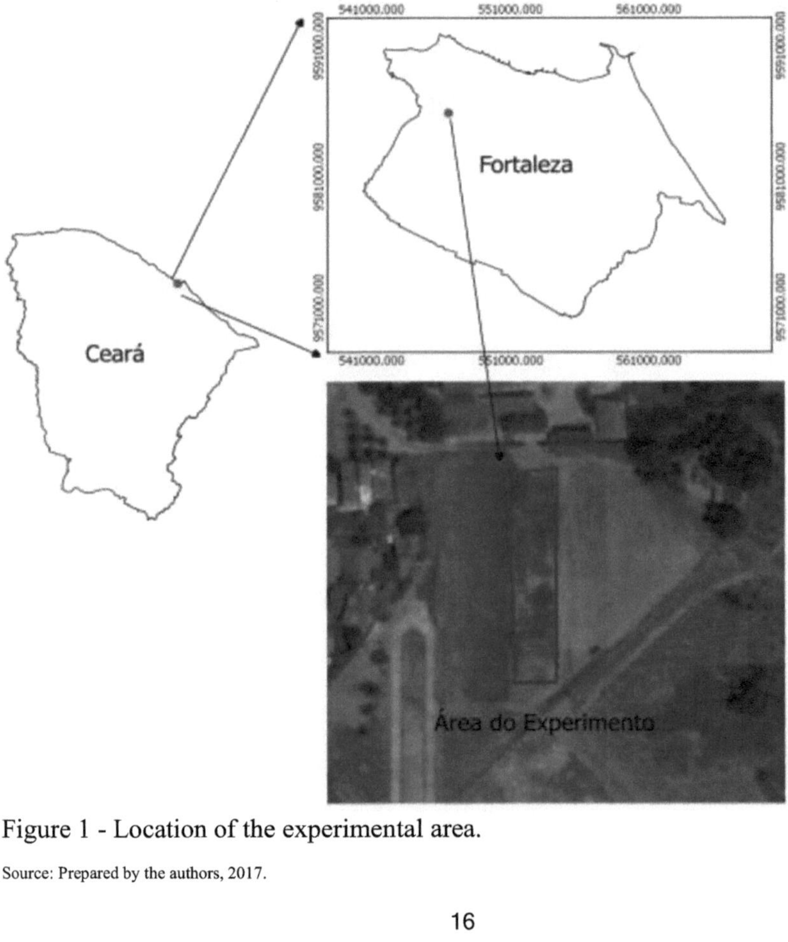

Figure 1 - Location of the experimental area.

Source: Prepared by the authors, 2017.

The experimental area is in the initial process of implementing the no-till system. In November 2014, crotalaria, sorghum and mombaça grass were planted to form mulch for sowing maize in March 2015. Following herbicide management, the bean crop was sown in August 2015. After the beans had been harvested and desiccated, Brachiaria brizantha was intercropped with maize between December 2015 and March 2016.

According to the Koppen classification (1923), the climate of the region where the experiment was carried out is defined as Aw', which indicates tropical rainy, very hot, with a predominance of rain in the summer and autumn seasons and an average temperature of over 18°C in all months. Figure 2 shows rainfall data, maximum and minimum temperatures and the average temperature for the period during which the experiment was conducted in the field, recorded at the Meteorological Station of the Department of Agricultural Engineering at UFC/ Fortaleza-CE.

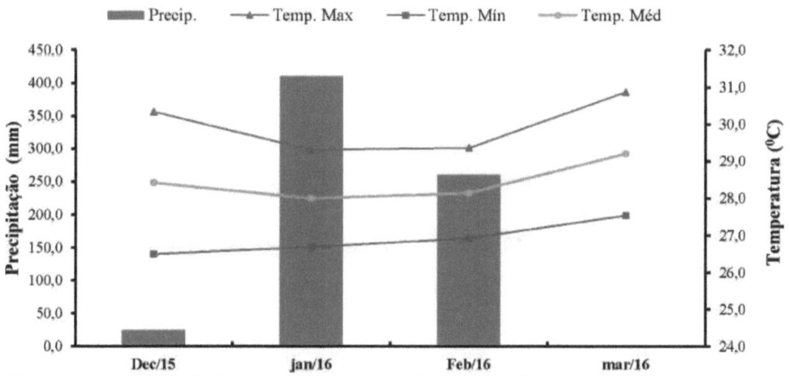

Figure 2 - Rainfall (mm), average maximum, minimum and average temperature (°C) during the experiment. Fortaleza, Ce.

Source: Prepared by the authors, 2017.

The soil in the experimental area was classified as Argissolo Vermelho Amarelo according to the EMBRAPA methodology (1999). Tables 1, 2 and 3 show the values obtained from the physical and chemical analysis of the soil carried out before the experiment was set up, with the soil sampled at depths of 0.0 - 0.20 and 0.20 - 0.40 metres.

Table 1- Soil granulometric analysis at different depths.

Layer	Clay	Silt	Sand		Class
			Thin	Thick	
	g kg^{-1}				
0 - 20	108	64	316	512	Free sand
20 - 40	147	51	296	506	Sandy loam

Source. Prepared by the authors.

Table 2 - Soil chemical characteristics assessed at different depths.

Layer	pH H2O	MO	P	K	Ca	Mg	H+Al	SB	T	V
		g dm^{-3}	mg dm^{-3}	mmolc dm^{-3}						%
0-20	6,8	4,65	55	2,4	14	9,0	14,9	25,4	40,3	63
20-40	5,7	4,55	13	1,8	19	5,0	16,5	25,8	42,3	61

MO: organic matter; P: phosphorus in resin; SB: sum of bases: T: cation exchange capacity at pH 7.0; V: soil base saturation.
Source. Prepared by the authors.

Table 3 - Physical characterisation of the soil in the experimental area at different depths.

Depth (m)	Macroporosity (m^3 m^{-3})	Microporosity (m^3 m^{-3})	Total Porosity (m^3 m^{-3})	Soil Density (kg dm^{-3})
0,0 - 0,1	0,15	0,09	0,24	1,64
0,1 - 0,2	0,14	0,09	0,23	1,66
0,2 - 0,3	0,14	0,08	0,22	1,66

Source. Prepared by the authors.

3.2 - Experimental design and description of treatments

A randomised block experimental design was used in a factorial scheme (2x3) + 2, with four replications, two corn spacings (0.45 and 0.9 m) and three forage sowing methods (pre-sowing corn, in the sowing row next to the base fertiliser, post-sowing corn) and single corn in both spacings as the control, totalling 32 experimental units. Each block had an area of 480 m^2 and consisted of eight plots (Figure 3). The plots with a spacing of 0.90 m were made up of three rows of maize, 3.0 m wide and 20 m long, while the plots with a spacing of 0.45 m were made up of six rows of maize. The useful area of the plot was considered to be the 10 m long centre row for the 0.90 m spacing and the two 10 m long centre rows for the 0.45 m spacing.

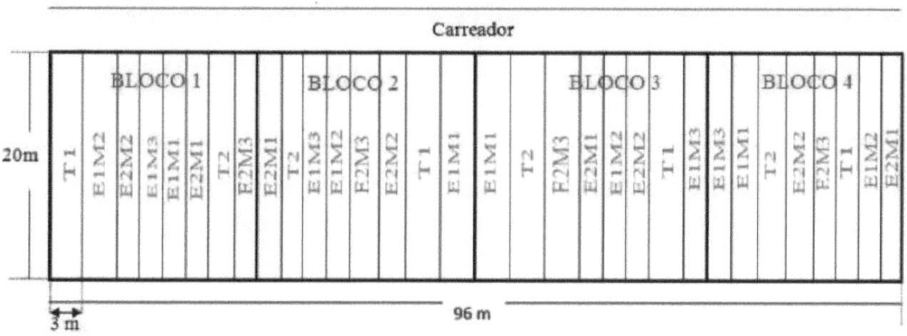

Figure 3 - Sketch of the experimental area.
Source. Prepared by the author.

In which,

E1 = Spacing 1 - 0.45 m;

E2 = Spacing 2 - 0.90 m;

M1 = Mode 1 - forage sown in the same row as the maize;

M2 = Mode 2 - Forage sown in the field before maize sowing;

M3 = Mode 3 - Forage sown in the field after maize sowing;

T1 = Test 1 - single maize with a spacing of 0.45 m between rows;

T2 = Witness 2 - single maize with a spacing of 0.90 m between rows.

3.3 - Conducting the experiment

3.3.1 - Irrigation

The experimental area received supplementary irrigation using the conventional sprinkler system, according to the water demand of the different phenological stages of the maize crop and rainfall in the growing months, with a shorter irrigation interval in the period of greatest water demand. ET0 was calculated using the class A tank method, obtaining the tank coefficients using the method proposed by Doorenbos & Pruitt (1977). When calculating the ETC, the ET0 was multiplied with the Kc of the different stages of the main crop (maize).

3.3.2 . Seeds

- Maize

Seeds of AG 1051 hybrid maize were used, aiming for a population of 60,000 plants ha^{-1}, with a sowing density of 8 seeds m^{-1} at a spacing of 0.90 metres.

m and 4 seeds m^{-1} for the 0.45 m spacing, taking into account germination values (85%), purity (93%), seedling survival (90%) and seeder slippage (5%).

- Fodder

The intercropping used 12 kg ha^{-1} of Brachiária brizantha cv. Marandu, with certified seeds and a cultural value of 50%.

According to Melotto et al. (2013) the seed density for Brachiaria brizantha cv. Marandu to provide a large amount of green matter when intercropped with maize can reach 600 VC points. Cultural value points are the quantity of seeds, in kg ha^{-1}, multiplied by their cultural value.

3.3.3 - Fertilisers

The NPK fertiliser applied at sowing time was 300 kg ha^{-1} of the commercial NPK formulation (10-28-20), established according to the results of the chemical analysis and soil fertility. For top dressing, urea and potassium chloride were used in the amount of 311 and 100 kg ha^{-1} respectively, split into three doses, the first being applied at the v2 stage, the second top dressing at the V4 stage and the third at the V6 stage of maize development. The amount of top dressing fertiliser was determined on the basis of the soil analysis and the maize's requirement to express its productive potential.

3.3.4 - Agricultural pesticides

Glyphosate was used in a total area, before sowing the maize, to desiccate the plants present in the experimental area and also to desiccate the Brachiaria 30 days after harvesting the maize.

To control the cartridge caterpillar ***(Spodoptera frugiperda),*** the product Lanatte (i.a) was used at a dose of 0.6 l ha^{-1} of the active ingredient.

3.3.5 - Machinery and equipment

The following were used to set up and conduct the experiment:

- 4x2 TDA (Auxiliary Front Wheel Drive) tractor, model VALTRA A950, with a maximum engine power of 70.6 kW (96 hp), fitted with 14.9 - 24 R-1 diagonal tyres on the front axle and 18.4 - 34 R-1 tyres on the rear axle. Used to pull the sprayer for herbicide application;
- 4x2 TDA (Auxiliary Front Wheel Drive) tractor, with 88.26 kW (120 hp), equipped with 14.9-24 R-1 diagonal tyres on the front axle and 18.4-34 R-1 diagonal tyres on the rear axle, air inflation pressure in the tyres of 110 and 124 kPa for the front and rear tyres respectively. Used in the seeding operation to drive the seeder;
- A mounted sprayer with a 14 m long boom, 28 110-02 fan tips spaced 0.50 m apart and a tank with a capacity of 600 litres of spray used to apply herbicides;
- Jumil precision pneumatic seeder, model JM2090 PD, with three rows, weighing approximately 1,160 kg, assembled, with a maximum capacity of 39 L in the fertiliser and seed tanks. Equipped with a furrowing rod for fertiliser deposition, a double mismatched disc for seed deposition, a vertical pneumatic seed dosing disc and a concave rubber wheel with central relief for seed compaction;
- 20 litre knapsack sprayer, used to apply insecticide;
- Nogueira model EN-9 F3B stationary ensilage machine, used to grind plant material for silage;
- Greenhouse, for drying plant material and soil;
- Scales, used for weighing plant material and soil;
- Tension table for determining soil porosity;
- Digital callipers to check the diameter of the stalks of maize plants.
- Rings and tins for collecting soil.

3.4 - Statistical data analysis

The data was subjected to descriptive analysis to determine the parameters symmetry and kurtosis, used to check the normality of the data according to the methodology proposed by Mesquita et al. (2003). After confirming the normality of the data, analysis of variance (ANOVA) was carried out to compare the means.

The operational performance evaluations were carried out in a 2x3 factorial scheme, considering spacing and sowing methods.

The factorial scheme (2x3) +1 was used to compare the control with the other treatments. The data was subjected to the F test and, when necessary, the Tukey test ($p<0.05$) was applied to compare the means of the consortia. The Dunnett test ($p<0.05$) was used to compare the factorial with the control (maize only).

3.5 - Determination of the soil's physical attributes

Samples were taken before the experiment was set up and after the maize was harvested in consortium with the forage plant. The data obtained before planting was used to characterise the chemistry and physics of the experimental area.

The following physical attributes of the soil were determined: macroporosity, microporosity and total porosity, using the tension table method recommended by Kiehl (1979). These evaluations used volumetric determinations on naturally structured monoliths with a known volume (100 cm^3), one point per plot, in the 0.0 - 0.10 m, 0.10 - 0.20 m and 0.20 - 0.30 m layers, using the methodology recommended by Embrapa (1997).

3.6 - Operational performance of the tractor-seeder combination

3.6.1 - Travel speed

The actual travel speed of the tractor-seeder combination was determined as a function of the space travelled in each experimental unit and the travel time taken by the tractor, obtained using a digital stopwatch. Equation 1.

$$V = \frac{\Delta S}{\Delta T} x\ 3{,}6 \qquad (1)$$

in which,

V = speed (km $h^{-1)}$;

ΔS = space travelled (m);

ΔT = time to travel the parcel (s);

3.6 = unit conversion factor.

3.6.2 - Operational field capacity

The operational field capacity was determined using the working width of the seeder, the actual travelling speed in km h^{-1} and the seeder's operating efficiency of 75% (Equation 2), according to ASAE (1997).

$$CCo = L\, x\, V\, x\ 0{,}36\, x\, 0{,}75 \qquad (2)$$

in which,

CCo − Operational field capacity (ha $h^{-1)}$;

L = Useful working width of the seed drill (m);

V = Take-off speed (km $h^{-1)}$;

0.75 = Operational efficiency;

0.36 = Unit conversion factor.

3.6.3 - Skating

Tractor wheel slip was determined using Equation 3.

$$PAT = \left[\frac{n^1 - n^0}{n^1}\right] x\ 100 \qquad (3)$$

in which,

PAT = Wheel slip, %;

n^0= Number of wheel turns without load;

n^1= Number of wheel turns under load.

The wheel slip of the seeder was determined using Equation 4.

$$PS= \left[\frac{(N \times Pr)-Le}{(N \times Pr)}\right] \times 100 \quad (4)$$

in which,

PS = Seeder slip (%);

N = Number of turns of the seeder drive wheel;

Pr = Perimeter of the seeder's drive wheel (m);

Le = Length of experimental line (m).

3.6.4 - Hourly fuel consumption

To determine fuel consumption, two **oval "Flowmate"** flow meters were used, **model Oval** M-III and LSF 41 with 0.01 ml accuracy, installed in series at the inlet and return of the injection pump (Figure 4). The pulses generated by the flowmeters are converted into volume, considering a flow rate of 1 ml $pulse^{-1}$. Fuel consumption was determined in the useful area of all the plots in units of volume (mL), and by the difference between the volumes of fuel measured in the injector pump feed pipe and the return pipe to the tank. The beginning and end of the useful area was demarcated using wooden stakes fixed in the ground 10 metres apart, which served as a reference point for switching the flow meters on and off.

Figure 4 - Flow meters installed on the tractor to quantify fuel consumption.

Source: Prepared by the authors, 2017.

3.6.5 - Fuel consumption per hectare

Fuel consumption per hectare was determined by the product of operational field capacity and hourly fuel consumption (Equation 5).

$$C\ (ha) = Ch\ x\ \mathrm{CCo} \qquad (5)$$

in which,

C (L ha^{-1}) = fuel consumption per hectare;

Ch = hourly consumption (L h^{-1});

CCo = operational field capacity (ha h^{-1}).

3.6.6 - Harvesting

As the material was intended for silage, it was harvested 85 days after sowing. All maize and brachiaria plants in the useful area of the plot were cut manually at 20 cm from the ground (Figure 5), at the R5 stage of the maize, which is characterised by the transition from the pasty to the mealy state of the grains, and also by having a dry matter (DM) content of 30 to 35% according to Carvalho, (2013).

Figure 5 - Representation of the cutting height of the plant material during harvesting.

Source: prepared by the authors, 2017.

CHAPTER 4

RESULTS AND DISCUSSION

Appendix 1 contains **the** symmetry and kurtosis **coefficients** for the variables evaluated. It can be seen that all the values obtained for these coefficients are within the range of -2 and 2. According to Montgomery (2004), values for the symmetry and kurtosis coefficients within the range of -2 and 2 indicate that the data follow a normal distribution, representing little deviation from the normal distribution, which can be considered the hypothesis of data normality, a necessary condition for carrying out the analysis of variance with confidence in obtaining the results.

For the soil macro and microporosity data (Table 4), it can be seen that most of the depths analysed showed no significant difference ($p<0.05$), depending on the spacing and sowing method factors.

Table 4 - Average values obtained for Macroporosity and Microporosity as a function of spacing and sowing methods.

Causes of Variation		Depths (m)			Depths (m)		
		0,0 - 0,10	0,10 - 0,20	0,20-0,30	0,0-0,10	0,10-0,20	0,20 - 0,30
		Macroporosity ($m^3\ m^{-3}$)			Microporosity ($m^3\ m^{-3}$)		
Spacing (E)	E1	0,09	0,07 b	0,07	0,14	0,15	0,15
	E2	0,11	0,12 a	0,08	0,13	0,13	0,13
Modality (M)	M1	0,11	0,12	0,08	0,14	0,15	0,15
	M2	0,11	0,08	0,09	0,13	0,14	0,14
	M3	0,08	0,09	0,06	0,14	0,13	0,14
F value	E	1.07 NS	4,67*	0.51 NS	0.32 NS	2.74 NS	1.26 NS
	M	0.90 NS	1.02 NS	0.70 NS	0.20 NS	0.41 NS	0.47 NS
	M*E	0.20 NS	1.75 NS	1.25NS	0.76 NS	1.72 NS	0.69 NS
DMS	E	0,04	0,04	0,03	0,02	0,03	0,02
	M	0,07	0,07	0,05	0,04	0,05	0,03
CV (%)		53,38	57,36	52,84	24,83	28,63	20,97

Averages without letters in the columns do not differ by Tukey's test at 5% probability. *- significant ($p<0.05$); NS- not significant ($p>0.05$). E1 - spacing 1 (0.45 m); E2 - spacing 2 (0.90 m). M1 - mode 1 (forage in the maize row); M2 - mode 2 (forage sown at the same time as pre-sowing maize); M3 - mode 3 (forage sown at the same time as post-sowing maize). MSD - minimum significant difference. CV - coefficient of variation.

Only at a depth of 0.10 - 0.20 did macroporosity show a significant difference, with a higher value at a spacing of 0.90 metres. This result may be associated with the fact that the smaller spacing provides better restructuring of the soil aggregates, due to the greater exploitation of the soil by the root system as a result of the more equidistant distribution of plants in the area, when compared to the 0.45 m spacing between rows. According to Pires et al. (1998), reducing the spacing between rows leads to greater exploitation of the soil by the roots.

The results found in this work disagree with Pantano (2003) who, when evaluating two spacings of maize intercropped with Urochloa in different modalities, obtained no significant difference in macroporosity values in the 0.0-0.30 m layer.

There was no significant difference between the sowing methods for the soil porosity data at any of the depths analysed (Table 4). According to Duarte and de Maria (2013), one of the important aspects of the no-till system is the expected improvement in soil quality. Forage plants, especially those of the Brachiaria and Panicum genera, have a very long and vigorous root system, with the capacity to promote soil restructuring. However, this effect takes place over the long term and the area where the experiment in this study was located is in the initial phase of implementing the SPD.

A similar behaviour was observed by Pantano (2003) who, in a study of the intercropping of maize with Urochloas in the Ilha Solteira region, found that the sowing methods did not affect the soil porosity values in any of the layers assessed.

Table 5 shows the soil macro and microporosity data at all the depths assessed, compared to single maize spaced at 0.45 m and all the treatments studied in this study.

It can be seen that there was no significant difference for any of the variables in any of the treatments used, nor did they differ from the control. According to Andrade and Stone (2009), an ideal soil for agricultural production should have a total porosity of close to 50% and a percentage distribution of 34% of macropores and 66% of micropores.

Table 5 - Summary of the analysis of variance and the test of means for soil macro and microporosity compared to the control at 0.45 m spacing.

Causes of Variation	Depths (m)			Depths (m)		
	0,0 -0,10	0,10 - 0,20	0,20 - 0,30	0,0 -0,10	0,10 - 0,20	0,20 - 0,30
Spacing Modality	Macroporosity ($m^3 m^{-3}$)			Microporosity ($m^3 m^{-3}$)		

E1	M1	0,11	0,07	0,05	0,15	0,18	0,17
E1	M2	0,09	0,08	0,10	0,13	0,14	0,14
E1	M3	0,06	0,06	0,06	0,13	0,15	0,14
E2	M1	0,07	0,17	0,10	0,18	0,11	0,13
E2	M2	0,08	0,07	0,08	0,14	0,15	0,13
E2	M3	0,06	0,12	0,7	0,15	0,11	0,14
T1		0,12	0,14	0,7	0,14	0,12	0,13
F-value - Factor x Control		3.30NS	2.63 NS	0.03 NS	0.14 NS	0.96 NS	0.16 NS
DMS		0,07	0,11	0,09	0,06	0,07	0,05
CV (%)		43,41	54,12	56,91	23,06	27,91	20,74

Averages followed by the same letter or without letters in the columns do not differ by the Dunnet test at 5% probability. NS- not significant (p>0.05). E1 - spacing 1 (0.45 m); E2 - spacing 2 (0.90 m). M1 - mode 1 (forage in the maize row); M2 - mode 2 (forage sown at the same time as pre-sowing maize); M3 - mode 3 (forage sown at the same time as post-sowing maize). MSD - minimum significant difference. CV - coefficient of variation.

It can be seen that after intercropping maize with forage (Tables 4, 5, 6 and 7) there was an increase in micro and a reduction in macroporosity and no change in total porosity compared to the initial soil physics analysis (Table 3). This result is due to better exploitation of the area by the roots, as discussed above.

Table 6 shows the soil macroporosity and microporosity data for all the depths assessed, compared to single maize spaced at 0.90 m, as well as all the treatments studied in this study. It can be seen that there was no significant difference for any of the variables in any of the treatments used, nor did they differ from the control.

When analysing sowing systems (row and inter-row) at depths of 0 to 0.10 m and 0.10 to 0.20 m in Eutroferric Red Latosol, Seidel et al. (2014) also found no significant difference in soil macroporosity and microporosity.

According to the authors, micropores are more resistant to deformation and therefore little altered by management in latosols.

Table 6 - Summary of the analysis of variance and the test of means for soil macro and microporosity compared to the control at 0.90 m spacing.

Causes of Variation		Depths (m)			Depths (m)		
		0,0 - 0,10	0,10 - 0,20	0,20 - 0,30	0,0 - 0,10	0,10 - 0,20	0,20 - 0,30
Spacing	Modality	Macroporosity ($m^3 m^{-3}$)			Microporosity ($m^3 m^{-3}$)		
E1	M1	0,11	0,07	0,05	0,15	0,18	0,17
E1	M2	0,09	0,08	0,10	0,13	0,14	0,14
E1	M3	0,06	0,06	0,06	0,13	0,15	0,14
E2	M1	0,11	0,17	0,10	0,12	0,11	0,13
E2	M2	0,12	0,07	0,08	0,12	0,15	0,13

E2	M3	0,10	0,12	0,07	0,15	0,11	0,14
T2		0,05	0,08	0,09	0,18	0,14	0,12
F-value - Factor x Control		2.40^{NS}	0.14^{NS}	0.16^{NS}	6.40^{NS}	0.02^{NS}	1.77^{NS}
DMS		0,10	0,11	0,08	0,07	0,07	0,05
CV (%)		54,53	59,56	51,16	24,74	27,58	20,48

Averages followed by the same letter or without letters in the columns do not differ by Tukey's test at 5% probability. ^{NS}- not significant ($p>0.05$). E1 - spacing 1 (0.45 m); E2 - spacing 2 (0.90 m). M1 - mode 1 (forage in the maize row); M2 - mode 2 (forage sown at the same time as pre-sowing maize); M3 - mode 3 (forage sown at the same time as post-sowing maize). MSD - minimum significant difference. CV - coefficient of variation.

For the total porosity values (Table 7) assessed in the 0.0-0.10, 0.10-0.20 and 0.20-0.30 m layers, there was no significant difference between the depths or between the sowing methods. The total porosity of the soil is made up of the sum of the macroporosity and microporosity values, which together make up 50 per cent of its volume, according to Kiehl (1979). The difference in the amount of macropores was not enough to affect the amount of total pores.

Silva et al. (2014), when assessing the behaviour of the physical attributes of a typical dystrophic Yellow Latosol, also found no change in the amount of total pores as a function of depth. However, Spera et al. (2010) observed that total porosity in the 0.0-0.05 m layer showed higher values when they assessed the soil's physical attributes after ten years of cultivation. According to Marcolan and Anghinoni (2006), as the years go by, the total porosity of no-till soil can partially increase as the soil structure improves.

Table 7 - Summary of the analysis of variance and the test of means for Total Soil Porosity as a function of spacing and sowing method.

Factor		Depths (m)		
		0,0 - 0,10	0,10 - 0,20	0,20 - 0,30
		Total Porosity ($m^3 m^{-3}$)		
Spacing (E)	E1	0,23	0,23	0,22
	E2	0,25	0,25	0,22
(C)	M1	0,25	0,27	0,23
	M2	0,24	0,22	0,23
	M3	0,22	0,23	0,20
F value	E	0.67 NS	0.87 NS	0.01 NS
	M	1.03^{NS}	1.44^{NS}	0.98^{NS}
	E*M	1.33^{NS}	0.15^{NS}	0.52^{NS}

DMS	E	0,03	0,04	0,03
	M	0,05	0,07	0,05
CV (%)		19,04	23,80	19,75

Averages without letters in the columns do not differ by Tukey's test at 5% probability. NS- not significant (p>0.05). E1 - spacing 1 (0.45 m); E2 - spacing 2 (0.90 m). M1 - modality 1 (forage in the maize row); M2 - modality 2 (forage sown at the same time as pre-sowing maize); M3 - modality 3 (forage sown at the same time as post-sowing maize). MSD - minimum significant difference. CV - coefficient of variation.

Table 8 shows the total porosity data at the three depths assessed, compared to single maize spaced at 0.45 m and 0.90 m, respectively, and with all sowing methods. There was also no significant difference, a result that may be associated with the fact that the no-till system is in its initial implementation phase and the benefits it brings to the physical quality of the soil have not yet been realised. According to Stone and Silveira (2001) and Oliveira et al. (2003) the physical attributes of the soil under no-till farming can be improved, however, the intensity of the improvement depends on the cultivation period, the number of crops per year and the species cultivated.Seidel et al. (2014), evaluating the physical properties of the soil after the intercropping of maize with brachiaria depending on the modalities and sowing times of the forage, observed greater total porosity when the brachiaria was sown in the inter-row system simultaneously with the maize. However, the area in which the experiment was carried out has been cultivated for over 20 years using the no-till system.

Table 8 - Summary of the analysis of variance and test of means for total soil porosity compared to the control at 0.45 m spacing (E1) and 0.90 m spacing (E2).

Causes of Variation		Depths (m)					
		0,0 - 0,10		0,10 - 0,20		0,20 - 0,30	
Spacing	Modality	E1	E2	E1	E2	E1	E2
		Total Porosity (m^3 m^{-3})					
E1	M1	0,27	0,27	0,25	0,25	0,23	0,23
E1	M2	0,23	0,23	0,22	0,22	0,24	0,24
E1	M3	0,20	0,20	0,21	0,21	0,20	0,20
E2	M1	0,25	0,24	0,29	0,29	0,24	0,24
E2	M2	0,22	0,25	0,23	0,23	0,21	0,22
E2	M3	0,21	0,25	0,23	0,23	0,21	0,21
T1 - T2		0,26	0,24	0,27	0,22	0,22	0,21
F-value - Factor x Control		1.22 NS	0.04 NS	0.75 NS	0.26 NS	0.23 NS	0.24 NS
DMS		0,10	0,08	0,12	0,10	0,08	0,08

CV (%)	21,56	17,83	25,28	23,07	19,38	19,73

Averages followed by the same letter or without letters in the columns do not differ by Dunnet's test at 5% probability NS - not significant ($p>0.05$). E1 - spacing 1 (0.45 m); E2 - spacing 2 (0.90 m). M1 - mode 1 (forage in the maize row); M2 - mode 2 (forage sown at the same time as pre-sowing maize); M3 - mode 3 (forage sown at the same time as post-sowing maize). MSD - minimum significant difference. CV - coefficient of variation.

Table 9 shows the results of the analysis of variance and the average values obtained for tractor front and rear wheel slip, seeder slip/sliding, fuel consumption and operational field capacity.

The slip of the tractor wheels, both front and rear, showed no significant difference for the variables analysed. However, the values observed are outside the ASAE (1989) standard, which recommends a slip of 8 to 10% for the tractor to operate with maximum traction efficiency on uncultivated soils. The slip values indicate that the tractor may have been oversized for the operation.

Garcia et al. (2006) state that it is important to know the machine's capacity in order to select the power and equipment that will carry out agricultural operations in a timely manner, avoiding additional costs with oversized or undersized machines, which are common on farms.

Table 9 - Average values obtained for the operational and energy performance of the tractor-seeder combination.

Causes of Variation		Skidding (%) RT	Skidding (%) RD	Slippage (%) Seeder	Fuel consumption (L h^{-1})
Spacing (E)	E1	2,62	3,65	-11,11 a	10,78
	E2	3,29	4,89	-24,07 b	11,41
Modality (M)	M1	1,86	4,57	-14,58	10,66
	M2	1,92	4,46	-17,36	11,07
	M3	4,08	3,78	-20,83	11,56
F value	E	0.84 NS	2.69 NS	25,56*	1.47 NS
	M	3.03 NS	0.43 NS	1.98 NS	0.98 NS
	E*M	0.48 NS	0.30 NS	0.22 NS	3,80 *
DMS	E	1,54	1,58	5,38	1,09
	M	2,30	2,35	8,01	1,63
CV (%)		61,05	43,2	35,70	11,53

Averages followed by the same letter or without letters in the columns do not differ by Tukey's test at 5% probability. *- significant ($p<0.05$); NS- not significant ($p>0.05$). RD - front wheelset; RT - rear wheelset. E1 - spacing 1 (0.45 m); E2 - spacing 2 (0.90 m). M1 - mode 1 (forage in the maize row); M2 - mode 2 (forage sown at the same time as maize pre-sowing); M3 - mode 3 (forage sown at the same time as maize post-sowing). MSD - minimum significant difference. CV - coefficient of variation.

For the slippage of the seeder wheels, there was a significant difference only for the spacing factor, with the lowest value at spacing 1 (0.45 m between rows). This result may be associated with the concentration and distribution of the seeder's load over the rows, due to the fact that at the lower spacing they are close together, improving wheel-soil contact and contributing to less wheel slippage.

According to Vieira and Reis (2001), the efficiency of seed drills with wheel-driven seed and fertiliser metering mechanisms is directly related to the conditions of wheel-soil contact and is directly influenced by the vertical load. They also state that the weight of the seeder influences wheel slip.

According to Kamphorst (2003), patination can cause variations of up to twenty per cent or more, resulting in fewer seeds per hectare than the required seed rate.

Fuel consumption in L h^{-1} was statistically different and the interaction between the factors was significant ($p<0.05$). Figure 6 shows the results of the breakdown of the interaction between the spacing and sowing method factors for hourly fuel consumption.

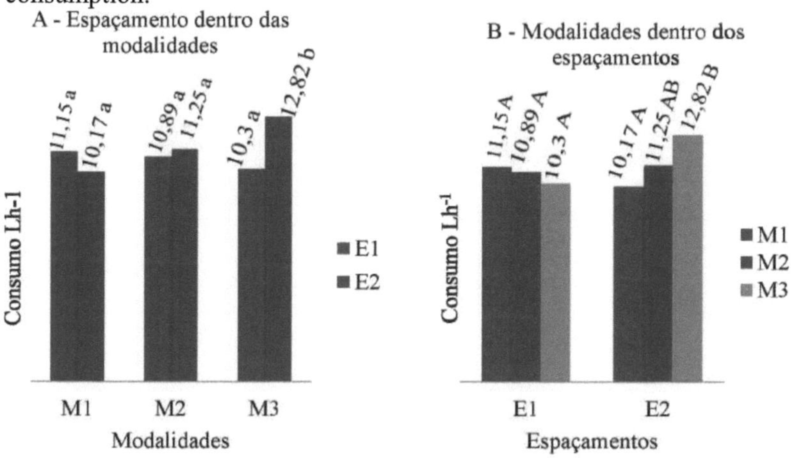

Fonte: Elaborada pelos autores, 2017.

Figure 6 - Breakdown of the significant interaction for hourly fuel consumption when sowing maize as a function of spacing and sowing method.

When splitting the spacings between the methods, there was a difference only in method 3, with the highest consumption value for spacing 2 (0.90 m between rows). When splitting the methods within the spacings, it can be seen that at spacing 1 (0.45 m between rows) there was no significant difference for any sowing method. At spacing 2, there was a significant difference between modality 1 (forage sown in the maize row) and modality 3 (forage sown at the same time as maize), with the highest hourly fuel consumption being observed in modality 3. However, the sowing mode factor may not have influenced the fuel consumption of the mechanised unit, taking into account that these treatments were still being implemented, as the forage in mode 3 would be sown after sowing the maize and could therefore not interfere with this variable. These results may be related to factors not analysed in this study, such as the conditions described by Mialhe (1996).

Table 10 shows the results of the analysis of variance and the average values obtained for fuel consumption per area, operational field capacity and speed of movement of the mechanised unit as a function of row spacing and sowing methods.

Table 10 - Average values obtained for fuel consumption per area, operational field capacity and travel speed of the mechanised unit.

Causes of Variation		Fuel consumption (L ha^{-1})	CCO (ha h^{-1})	Vel. (km h^{-1})
Spacing (E)	E1	9,23	1,16	5,75
	E2	9,54	1,19	5,90
Modality (M)	M1	9,06	1,17	5,81
	M2	9,35	1,18	5,84
	M3	9,75	1,17	5,82
F value	E	0.55 ^{NS}	2.27 ^{NS}	2.27 ^{NS}
	M	0.90 ^{NS}	0.03 ^{NS}	0.03 ^{NS}
	E*M	2.12 ^{NS}	2.88 ^{NS}	2.88 ^{NS}
DMS	E	0,88	0,04	0,19
	M	1,31	0,06	0,29
CV (%)		10,98	3,98	3,98

Averages followed by the same letter or without letters in the columns do not differ by Tukey's test at 5% probability. *- significant ($p<0.05$); ^{NS}- not significant ($p>0.05$). RD - front wheelset; RT - rear wheelset. E1 - spacing 1 (0.45 m); E2 - spacing 2 (0.90 m). M1 - mode 1 (forage in the maize row); M2 - mode 2 (forage sown at the same time as pre-sowing maize); M3 - mode 3 (forage sown at the same time as post-sowing maize). MSD- minimum significant difference. CV - coefficient of variation; CCO - Operational Field Capacity.

The operational field capacity showed no difference between the parameters evaluated, explained by the use of the same travelling speed of the tractor-seeder combination, as well as the same working width in all the plots. The fuel consumption per area also showed no statistical difference, a result that can be associated with the OCC, since it is obtained by dividing the hourly consumption by the OCC, and this did not differ statistically between the treatments, as explained above.

Chioderoli et al. (2013), evaluating the performance of agricultural tractors in four types of maize intercropping with Urocloas plus the control, also found no influence of the treatments on the operating parameters. This result was considered very important, as changes in these parameters can directly affect operating costs.

Table 10 also shows that travel speed does not differ significantly ($p<0.05$) between the variables analysed, a result justified by the use of the same working gear and engine speed. This corroborates the results obtained by Queiroz (2015), who assessed the operational performance of the tractor-seeder combination when intercropping maize with Urochloas, working with the same spacing as this study and observing no change in travel speed.

CHAPTER 5

CONCLUSIONS

The 0.45 m spacing provided better structuring of the soil porosity when compared to the 0.90 m spacing.

The operational performance of the tractor-seeder combination showed better results at the 0.45 m spacing, with less slippage of the seeder wheels and better longitudinal seed distribution when compared to the 0.90 m spacing.

The 0.90 metre spacing associated with the forage seed sown after maize sowing provides greater fuel consumption.

REFERENCES

AMERICAN SOCIETY OF AGRICULTURAL ENGINEERS (ASAE). ASAE EP 497.6 JUN09: Agricultural machinery management data. In: **ASAE Standards 2009**: standards engineering practices data. St. Joseph, 2009. p.350-357.

AMERICAN SOCIETY OF AGRICULTURAL ENGINEERS (ASAE). Agricultural Machinery Management. EP 496. 2. In: **ASAE standards: Standards engineering practices data.** St. Joseph: American Society of Agricultural Engineers, 1997. p. 353-358.

AMERICAN SOCIETY OF AGRICULTURAL ENGINEERS. Terminology for soil-engaging components for conservation-tillage planters, drills and seeders. In:. **ASAE standards 1996: standards engineering practices data.** San Joseph, 1996. p. 309-314. (ASAE D477. 2 DEC93).

AMERICAN SOCIETY OF AGRICULTURAL ENGINEERS (ASAE). Agricultural tractor test code. In: **ASAE standards 1989: standards engineering practices data**. St. Joseph, 1989. p. 44-8.

AMERICAN SOCIETY OF AGRICULTURAL ENGINEERS (ASAE), S296. **Uniform termonology for traction of agricultural tractors, self - propelled implements, and other traction and transport devices.** St. Joseph, 1982.

ANDRADE, R. S.; STONE, L. F. S index as an indicator of soil physical quality. Revista Brasileira de Engenharia Agrícola e Ambiental, Campina Grande, PB, V.13, n.4, p.382-388, 2009.

BALASTREIRE, L. A. **Agricultural machinery**. São Paulo: Manole, 1990. 310 p

BALASTREIRE, L. A. **Agricultural machinery.** São Paulo: Manole, 1987. 307 p.

BATEY, T. Soil compaction and soil management: a review. **Soil Use and Management,** Oxford, v. 25, p. 335-345, 2009.

BLANCO-CANQUI, H.; CLAASSEN, M. M.; STONE, L. R. Controlled traffic impacts on physical and hydraulic properties in an intensively cropped no-till soil. **Soil Science Society of America Journal.** Madison, v.74, p. 2142-2150, 2010.

BORSATTO, E. A. **Soil preparation systems and maize (Zea mays L.) crop development at two row spacings**. 2005. 63f. Dissertation (Master's Degree in Soil Science) - Faculty of Agricultural and Veterinary Sciences, Jaboticabal, São Paulo State University, 2005.

CAMARA, R. K.; KLEIN, V. A. Physico-hydric properties of soil under scarified no-till and soya yield. **Ciência Rural,** Santa Maria, v. 35, p. 813-819, 2005.

CARVALHO, I. Q. **Cutting Point of Maize for Silage.** abc Foundation. Available at:< http://www.fundacaoabc.org.br/forragicultura/banco_forragens/Ponto_Corte_Silagem Milh_o>. pdf Accessed on 16/12/2015.

CEPIK, C. T. C.; TREIN, C. R.; LEVIEN, R; CONTE, O. Tractive force and soil mobilisation by the furrowing shanks of combine harvesters. **Revista Brasileira Eng. Agríc. Ambiental,** v.14, n.5, p.561-566, 2010.

CHIODEROLI C. A. **Intercropping Urochloas with maize in a no-till system as a predecessor crop to summer soya.** 2013. 174p. Thesis (PhD) - Universidade Estadual Paulista, Faculdade de Ciências Agrárias e Veterinárias, Jaboticabal, 2013.

CHIODEROLI, C.; MELLO, L. M. M.; GRIGOLLI, P. J.; FURLANI, C. E. A.; SILVA, J. O. R.; CESARIN, A. L. Soil physical attributes and soya productivity in a maize and brachiara intercropping system. **Revista Brasileira de Engenharia Agrícola e Ambiental**, Campina Grande, v. 16, n. 1, p. 37-43. 2012.

CHIODEROLI, C. A.; FURLANI,C. E. A.; SILVA, R. P.; Gitti, D. C.; KANEKO, F. H.; ROMAN, R. A. A . Seeder-advocate performance as a function of soil preparation and maize crop spacing. **Pesquisa Agropecuária Tropical** (Impresso), v. 40, p. 462-467, 2010.

CONTE, O.; FLORES, J. P. C.; CASSOL, L. C.; ANGHINONI, I.; CARVALHO, P. C. F.; LEVIEN, R.; WESP, C. L. Evolution of soil physical attributes in a crop-livestock integration system. **Pesquisa Agropecuária Brasileira**, Brasília, v. 46, p. 1301-1309, 2011.

CORTEZ, J. W. **Sowing density of soya and depth of fertiliser deposition in the no-till system**. 2007. 87f. Dissertation (Master's degree in Plant Production) - Universidade Estadual Paulista, Faculdade de Ciências Agrárias e Veterinárias - FCAV, Jaboticabal.

DEFOSSEZ, P.; RICHARD, G. Models of soil compaction due to traffic and their evaluation. **Soil and Tillage Research**, Amsterdam, v. 67, p. 41-64, 2002.

DENARDIN, J. E.; KOCHHANN, R. A.; SANTI, A.; FAGANELLO, A.; SATTLER, A. **Effect of maize-brachiaria (Brachiaria brizantha) intercropping on soil compaction mitigation.** Passo Fundo: Embrapa Trigo, 2008.13 p. (Embrapa Trigo. Boletim de Pesquisa e Desenvolvimento, 54). Available at: <http://www. cnpt.embrapa.br/biblio/bp/p_bp54.htm>. Accessed on: 20 Dec. 2016.

DOORENBOS, J.; PRUITT, W. O. **Crop water requirements. Rome**: FAO, 1977 (Irrigation and drainage paper, 24).

Duarte, A. P. D.; de Maria, I. C. Maize + brachiaria: minimum investment, maximum return. A Lavoura, n. 697, p. 39 - 43, 2013.

EMBRAPA, National Soil Research Centre (Rio de Janeiro, RJ). **Brazilian Soil Classification System.** Rio de Janeiro, 1999, 412p.

FILHO, A. G.; LANÇAS, K. P.; LEITE, F.; ACOSTA, J. J. B.; JESUINO, P. R. Agricultural tractor performance on three soil surfaces and four travel speeds. **Revista Brasileira de Engenharia Agrícola e Ambiental**, v.14, n.3, p.333-339, 2010.

FLORES, J. P. C.; ANGHINONI, I.; CASSOL, L. C.; CARVALHO, P. C. de F.; J. LEITE, G. D. B.; & FRAGA, T. I. Soil physical attributes and soya yield in a no-tillage system in crop-livestock integration with different grazing pressures. **Revista Brasileira de Ciências do Solo**, Viçosa, 31:771- 780, 2007.

FURLANI, C. E. A.; GAMERO, C. A.; LEVIEN, R.; LOPES, A.; SILVA, R. P. Operational performance of a precision seeder-adviser as a function of soil preparation and winter cover management. **Engenharia Agrícola**, Jaboticabal, v. 24, n. 2, p. 388-395, 2004.

GARCIA, L. C. Influence of travelling speed on maize sowing. **Engenharia Agrícola**, Jaboticabal, v.26, n.2, p.520-527, 2006.

GENRO JUNIOR, S. A.; REINERT, D. J.; REICHERT, J. M.; ALBUQUERQUE, J. A. Physical attributes of a Red Latosol and productivity of crops grown in succession and rotation. **Ciência Rural**, Santa Maria, v. 39, p. 65-73, 2009.

GRISSO, R. D.; HOFMAN, V. L. **Pesticide application equipment**. In: Conservation tillage systems and management: crop residue management with no-till, ridge-till, mulch-till. Ames: Midwest Plan Service, 1992. p. 111- 115.

IMHOFF, S.; SILVA, A. P. da; DIAS JUNIOR, M. S.; TORMENA, C. A. Quantification of critical pressures for plant growth. **Revista Brasileira de Ciência do Solo**, Viçosa, v. 25, p. 11-18, 2001.

JASPER, S. P.; SILVA, P. R. A. Economic comparison of different furrowing mechanisms for no-till maize seed drills. **Nucleus**, v.12,n.1, 2015.

KAMPHORST, J. S. How to plant well. **Cultivar magazine**. 23 ed., p.1-15, 2003.

KIEHL, E. J. Manual de edafologia - relações solo-planta. São Paulo, **Agronômica Ceres**,

1979. 264p.

KOPPEN, W. **Die klimate der Erde**. Berlin: Guyter, 1923, 369 p.

KLUTHCOUSKI, J.; YOKOYAMA, L. P. **Crop-livestock integration options.** In: KLUTHCOUSKI, J.; STONE, L. F.; AIDAR, H. Crop-livestock integration. Embrapa Rice and Beans. Santo Antônio de Goiás: Embrapa Arroz e Feijão, 2003. p. 131-141.

LEVIEN, R.; GAMERO, C. A.; FURLANI, C. E. A. Mechanical management of black oats with a knife roller and tractorised straw shredder. In: BRAZILIAN CONGRESS OF AGRICULTURAL ENGINEERING, 30, 2001, Foz do Iguaçu. **Proceedings**... Brazilian Society of Agricultural Engineering, 2001. 1 CD ROM.

LEVIEN, R.; MARQUES, J. P.; BENEZ, S. H. Performance of a precision seeder-adviser when sowing maize (Zea mays L.) under different forms of soil management. In: BRAZILIAN CONGRESS OF AGRICULTURAL ENGINEERING, 28th, 1999, Pelotas. **Proceedings**...
Pelotas: Brazilian Society of Agricultural Engineering, 1999. 1 CD-ROM.

LIMA, H. V.; LIMA, C. L. R.; LEÃO, T. P.; COOPER, M.; SILVA, A. P. & ROMERO, R. E. Agricultural machinery traffic and changes in biopores in an area under orange orchard. **Revista Brasileira de Ciências do Solo**, 29:677-684, 2005.

LOSS, A.; PEREIRA, M. G.; GIÁCOMO, S. G.; PERIN, A.; ANJOS, L. H. C. dos. Aggregation, carbon and nitrogen in soil aggregates under no-till farming with crop-livestock integration. **Pesquisa Agropecuária Brasileira**, Brasília, v. 46, n. 10, p. 1269-1276, 2011.

MAHL, D. **Seeder performance as a function of cutting mechanism, speed and soil, in the no-till maize system**. 2006. 143 f. Thesis (Doctorate in Energy in Agriculture)-Faculty of Agronomic Sciences, São Paulo State University, Botucatu, 2006.

MAHL, D.; SANTOS, F. A.; YANO, E. H.; SILVA, P. R. A.; GAMERO, C. A. Influence of increased speed in the sowing operation of the lupin crop in a no-till system. In: CONGRESO ARGENTINO DE INGENIERÍA RURAL, 8., 2005, Villa de

Merlo. **Proceedings...** San Luiz, 2005. 1 CD-ROM.

MAHL, D.; GAMERO, C.A.; BENEZ, S.H.; FURLANI, C.E.A.; SILVA, A.R.B. Energy demand and efficiency in maize seed distribution under speed variation and soil condition. Engenharia Agrícola, Jaboticabal, v.24, n.1, p.150-7, 2004.

MAHL, D. **Performance of maize (Zea mays L.) seed drills in a no-till system.** 2002. 160 f. Dissertation (Master's Degree in Agronomy/Energy in Agriculture)-Faculty of Agronomic Sciences, São Paulo State University, Botucatu, 2002.

MARCHÃO, R. L.; BALBINO, L. C.; SILVA, E. M.; SANTOS JUNIOR, J. D. G.; SÁ, M. A. C. de; VILELA, L.; BECQUER, T. Physical quality of a Red Latosol under crop-livestock integration systems in the Cerrado. **Pesquisa Agropecuária Brasileira**, Brasília, v. 42, n. 6, p. 873-882, 2007.

MARCOLAN, A.L.; ANGHINONI, I. Physical attributes of an Argissolo and crop yield according to soil tillage in no-till. **Revista Brasileira de Ciência do Solo**, v.23, p.163-170, 2006.

MAZETTO, F. R.; LANÇAS, K. P.; NAGAOKA, A. K.; CASTRO NETO, P.; GUERRA, S. P. S. Evaluation of tyre-soil contact in three models of agricultural tyres. **Engenharia Agrícola**, Sorocaba, v.24, n.3, p.750-757, 2004.

MELOTTO, A. M.; LOURENÇÃO, A. L. F.; GRIGOLLI, J. F. J.; ROSCOE, R. Implementation of the Maize and Grass Consortium in Mato Grosso do Sul: Main Aspects. IN: Technology and Production, MS Foundation. Maize off-season and winter crops, 2013. p. 130-157.

MENDONÇA, V. Z.; MELLO, L. M.; ANDREOTTI, M.; PEREIRA, F. C. B. L.; LIMA, R. C.; VALÉRIO FILHO. W. V.; YANO, E. H. Evaluation of soil physical attributes in a consortium of forage crops and maize in succession with soya beans in a cerrado region. **Rev. Bras. Ciências do Solo**, Viçosa, v. 37, n.1, p. 251-259, 2013.

MESQUITA, M. G. B. F.; MORAES, O.; CORRENTE, J. E. 2003. Statistical characterisation of soil physical variables. **Acta Scientiarum Agronomy**, v. 25, n.1, p. 35-

44, 2003.

MIALHE, L. G. Manual of agricultural mechanisation. São Paulo: **Agronômica Ceres**, 1974. 301 p.

MIALHE, L. G. Testing & certification of tractors. In: MIALHE, L. G. **Máquinas agrícolas:** ensaio & certificação. Piracicaba: Luiz de Queiroz Agricultural Studies Foundation, 1996, p. 385-462.

MONTANHA, G. K.; GUERRA, S. P. S.; SANCHEZ, P. A.; CAMPOS, F. H.; LANÇAS, K. P. Fuel consumption of an agricultural tractor in soil preparation for irrigated cotton as a function of tyre inflation pressure. **Revista Energia na Agricultura**, Botucatu, v.26, n.1, 2011.

MONTGOMERY D. C, 2004. **Introduction to statistical quality control**. Rio de Janeiro. 563 p.

MOREIRA, W. H.; BETIOLI JUNIOR, E.; PETEAN, L. P.; TORMENA, C. A.; SÉRGIO JOSÉ ALVES, S. A.; MARCO AURÉLIO TEIXEIRA COSTA, M. A. T.; FRANCO, H. H. S. Physical attributes of a dystrophic Latossolo Vermelho in a crop-livestock integration system. **Rev. Bras. Ciênc. Solo**, Viçosa, v. 36 n.2, 2012.

NAGAOKA, A. K.; NOMURA, R. H. C. Tractors: sowing. **Cultivar Máquinas**, Pelotas, v. 18, n. 1, p. 24- 26, mar. 2003.

OLIVEIRA, G. C. de; DIAS JUNIOR, M. S.; RESK, D. V. S.; CURI, N. Alterations structural and compressive behaviour of a clayey dystrophic Red Latosol under different use and management systems. **Pesquisa Agropecuária Brasileira**, Brasília, v. 38, n. 2, p. 291-299, 2003.

OLIVEIRA, M. D. M. **Operating cost and renewal point of agricultural tyre tractors: evaluation of a fleet.** Escola Superior de *Agricultur*- "Luiz de Queiroz", ESALQ - Piracicaba, p.150. 2000. (Master's dissertation).

OLIVEIRA, M. L. **Evaluation of the performance of a no-till seeder-adviser in two soil classes with different types of vegetation cover**. 1997. 50 f. Dissertation (Master's Degree in

Agricultural Engineering)-Federal University of Viçosa, Viçosa, 1997.

PANTANO, A. C. **Sowing brachiaria in consortium with maize at different spacings in the integration of agriculture and livestock in no-till farming.** 2003.60 f. Dissertation (Master's Degree in Agronomy/Production Systems)- Faculty of Engineering- Paulista State University, Ilha Solteira, 2003.

PIRES, J.L.; COSTA, J.A.; THOMAS, A. L. Soya bean yield influenced by plant arrangement and fertiliser levels. Pesquisa Agropecuária Gaúcha, Porto Alegre, v.4, n.2, p.183-188, 1998.

QUEIROZ, R. F. **Consortium of maize with Urochloa ruziziensis at two spacings and sowing patterns.** 2015. 89 f. Dissertation (Master's in Agricultural Engineering) - Centre for Agricultural Sciences, Federal University of Ceará, Fortaleza, 2015.

RIBEIRO, M. F. S. Selection and management aspects of animal-drawn no-till seed drills. In: PEIXOTO, R. T. G.; AHRENS, D. C.; SAMAHA, M. J. No-till farming: the road to sustainable agriculture. Ponta Grossa: Agronomic Institute of Paraná, 1997. p. 252-256.

RODRIGUES, J. G. L.; FERNANDES, J. C.; NASCIMENTO, F. M.; GAMERO, C. A.; BICUDO, S. J. Physical characterisation of the soil and operational performance of agricultural machinery in the establishment of forage sorghum crops. **Semina: Ciências Agrárias**, Londrina, v. 32, suplemento 1, p. 1813-1824, 2011.

RODRIGUES, J. G. L.; GAMERO, C. A. Energy demand and effective field capacity in different soil management systems. **Energia na Agricultura**, Botucatu, v. 21, n. 4, p. 55-62, 2006.

ROSA, H. A.; VELOSO, G.; SECCO, D.; RUFFATO, G.G.; MARCONDES, L. C.; SANTOS, R. F. Effects of the time of adoption of the no-till system on the physical attributes of a clayey Latosol. **Revista Cultivando o Saber**, Cascavel, v.5, n.2, p. 165-174, 2012.

SALES, L. E. O.; CARNEIRO, M. A. C.; SEVERIANO, E. C.; OLIVEIRA, G. C.;

FERREIRA, M. M. Physical quality of Neossolo Quartzarênico submitted to different agricultural use systems. **Ciência e Agrotecnologia**, Lavras, v. 34, n. 3, p. 667-674, 2010.

SANTOS, H. P. DOS.; FONTANELI, R. S.; SPERA, S. T.; TOMM. G. O. Effect of crop-livestock integration (CLI) production systems on soil fertility in no-till. Acta Scientiarum Agronomy. v. 31 n. 4 Maringá, 2009.

SEIDEL, P. E.; GERHARDT, S. F. I.; CASTAGNARA.; NERES. A. D. M. Effect of time and sowing system of Brachiaria brizantha in consortium with maize on production components and soil physical properties. **Ciências Agrárias**, Londrina, v. 35, n. 1, p. 55-66, 2014.

SIMÕES, D.; SILVA, M. R.; FENNER, P. T. Operational performance and costs of subsoiling in eucalyptus plantations. **Biosci. J.**, Uberlândia, v. 27, n. 5, p. 692-700, 2011.

SILVA, R. L.; MATIAS, S. S. R.; LOBATO, M. G. R.; NÓBREGA, J. C. A. Soil physical attributes in different vegetation covers in the southern region of Piauí. **Revista Caatinga**, v. 27, n.3, 2014.

SILVA, J. V.; CAMARGO, R.; WENDLING, B.; PIRES, S. C. Crop-livestock integration under a no-till system in the Brazilian cerrado. **Enciclopédia biosfera**, Centro Científico Conhecer - Goiânia, v.7, N.12, p. 1-12, 2011.

SILVA, S. L. **Evaluation of no-till seed drills: Energy demand, longitudinal distribution and depth of seed deposition at different travelling speeds**. 2000. 125f. Thesis (Doctorate in Agronomy/Energy in Agriculture) - Faculty of Agronomic Sciences, São Paulo State University, Botucatu, 2000.

SILVEIRA, J. C. M.; FERNANDES, H. C.; Modolo, A. J.; Silva, S. L.; Trogello, E. Energy demand of a seeder-adviser at different travel speeds and engine speeds. **Revista Ciência Agronômica**, Fortaleza, v. 44, n. 1, p. 44-52, 2013.

SILVEIRA, J. C. M. **Operational performance of a tractor-seeder combination as a function of engine shaft speed and rotation.** Thesis, PhD in Agricultural Engineering, Federal University of Viçosa. 145 f, Viçosa, 2008.

SILVEIRA, G. M.; YANAI, K.; KURACHI, S. A. H Determining the field efficiency of conventional soil preparation, sowing and cultivation machinery. **Rev. bras. eng. agríc. Ambiental**, Campina Grande, v.10 no.1, p.220-224, 2006.

SIQUEIRA, R.; CASÃO JUNIOR, R. **Grain and oilseed cultivation worker: machines for handling cover crops and sowing in the no-till system**. Curitiba: SENAR-PR, 2004. 88 p.

SOIL SCIENCE SOCIETY OF AMERICA. Glossary of soil science terms. Available at: <http://www.soils.org/sssagloss/>. Accessed on 20 December 2015.

SPERA, S. T.; SANTOS, H. P.; FONTANELI, R. S. & TOMM, G. O. Effect of crop-livestock integration under no-tillage on some soil physical attributes after ten years. **Bragantia**, Campinas,p.695-704, 2010.

STONE, L. F. & GUIMARÃES, C. M. **Influence of crop rotation systems on soil physical attributes**. Santo Antônio de Goiás, Embrapa Arroz e Feijão, 2005. 15p. (Research and Development Bulletin).

STONE, L.F.; MOREIRA, J.A.A.; KLUTHCOUSKI, J. **Influence of pastures on the improvement of soil physical and water attributes.** In: KLUTHCOUSKI, J.; STONE, L.F.; AIDAR, H. Crop-livestock integration. Santo Antônio de Goiás: Embrapa, 2003. p.173-181.

STONE, L. F.; GUIMARÃES, C. M.; MOREIRA, J. A. A. Soil compaction in bean cultivation - Effects on soil physical and hydric properties. **Rev. Bras. de Engenharia Agrícola e Ambiental**, v. 6, n. 2, p. 207-212, 2002.

STONE, L. F.; SILVEIRA, P. M. da. Effects of tillage system and crop rotation on soil porosity and density. **Revista Brasileira de Ciência do Solo**, Viçosa, v. 25, n. 2, p. 395-401, 2001.

VALE, W. G. **Operational and energy performance of an agricultural tractor during mowing, ploughing and sowing operations**. Thesis, PhD in Plant Production, Universidade Estadual do Norte Fluminense. 217 f, Goytacazes, 2011.

VALE, W. G.; GARCIA, R. F.; THIEBAUT, J. T. L.; AMIM, R. T.; TOURINO, M.

C. C. Performance and sample sizing for the evaluation of a seeder-soiler in no-till and conventional tillage. **Acta Scientiarum.** Agronomy, v. 30, n. 4, p. 441-448, 2008.

VIEIRA, L.B., REIS, E.F. **Machines for no-till farming.** Inf. Agropecu., v.22, n.208, p.43-8, 2001.

APPENDIX A - ASYMMETRY AND KURTOSIS RESULTS FOR ALL PARAMETERS EVALUATED

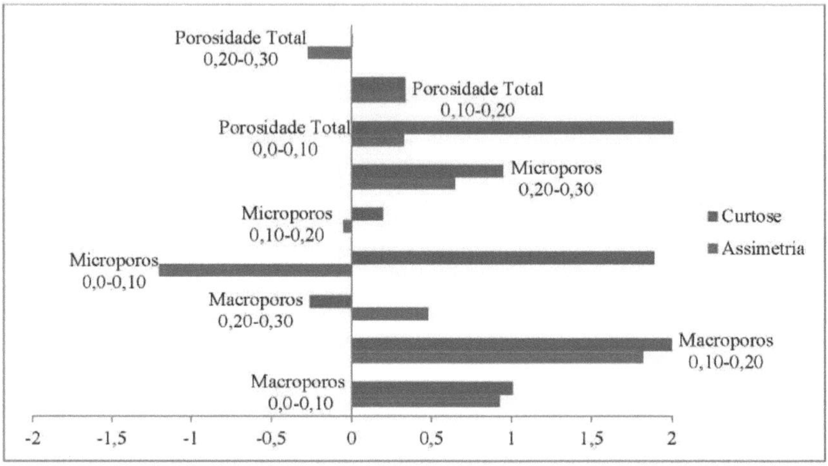

Figura 1. Asymmetry and kurtosis of the macroporosity, microporosity and total porosity data of the soil evaluated in the factorial scheme (3x2), as a function of spacing and sowing method.

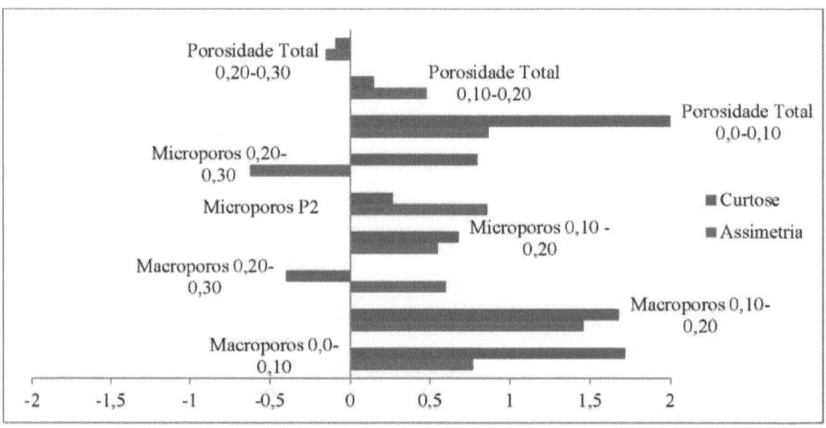

Figura 2. Asymmetry and kurtosis of the macroporosity, microporosity and total

porosity data of the soil evaluated in the factorial scheme (3x2) + 1, considering T1 - 0.45 m spacing.

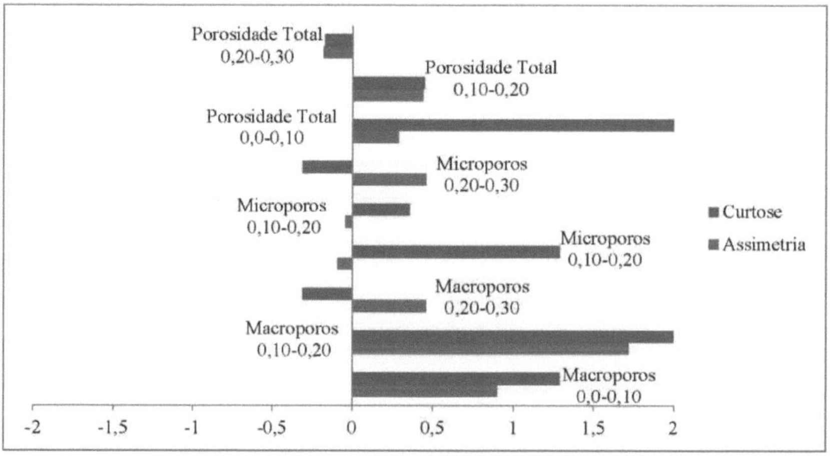

Figura 3. Asymmetry and kurtosis of the macroporosity, microporosity and total porosity data of the soil evaluated in the factorial scheme (3x2) + 1, considering T2 - 0.90 m spacing.

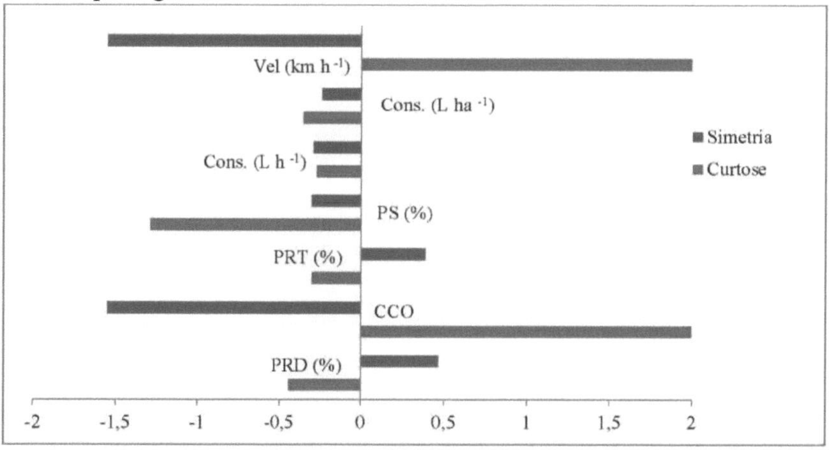

Figura 4. Coefficients of asymmetry and kurtosis for the parameters: Front wheel slip; Rear wheel slip; Seeder slip; Fuel consumption in L.ha^{-1}; Fuel consumption in L h^{-1}; Travel speed and Operational field capacity.

MIX
Papier aus verantwortungsvollen Quellen
Paper from responsible sources
FSC® C105338

Printed by Books on Demand GmbH, Norderstedt / Germany